11+

Non-Verbal
Reasoning

Practice
Papers

1

Although every effort has been made to ensure that website addresses are correct at time of going to press, Galore Park cannot be held responsible for the content of any website mentioned in this book. It is sometimes possible to find a relocated web page by typing in the address of the home page for a website in the URL window of your browser.

Hachette UK's policy is to use papers that are natural, renewable and recyclable products and made from wood grown in well-managed forests and other controlled sources. The logging and manufacturing processes are expected to conform to the environmental regulations of the country of origin.

Orders: Teachers, please contact Hachette UK Distribution, Hely Hutchinson Centre, Milton Road, Didcot, Oxfordshire OX11 7HH. Telephone: (44) 01235 400555. Email: primary@hachette.co.uk. Lines are open from 9 a.m. to 5 p.m., Monday to Friday.

Parents, Tutors please call: (44) 02031 226405 (Monday to Friday, 9:30 a.m. to 4.30 p.m.). Email: parentenquiries@galorepark.co.uk

Visit our website at www.galorepark.co.uk for details of other revision guides for Common Entrance, examination papers and Galore Park publications.

ISBN: 978 1 471849 30 5

© Hodder & Stoughton 2016
First published in 2016 by
Hodder & Stoughton Limited
An Hachette UK Company
Carmelite House
50 Victoria Embankment
London EC4Y 0DZ

Impression number   14 13 12
Year   2025 2024

Typeset in India

Illustrations by Peter Francis

Printed in the UK

A catalogue record for this title is available from the British Library.

www.carbonbalancedprint.com
CBP2250

Name: _____

11+

Non-Verbal Reasoning

Practice Papers

1

Peter Francis, Neil R. Williams & Sarah Collins

GALORE PARK

AN HACHETTE UK COMPANY

Contents and progress record

Practice Papers 1 contains papers
1–8 and should be attempted first

Ideal for...	Paper	Page	Length (no. Qs)	Timing (mins)
General training for all 11+ and pre-tests, good for familiarisation of test conditions.	1	9	20	13:30
	2	15	20	13:30
	3	19	20	10:00
	4	24	20	13:30
Short-style tests designed to increase speed. Good practice for pre-test and CEM 11+.	5	28	20	10:00
	6	33	20	13:30
	7	37	30	11:00
	8	44	10	5:00

Practice Papers 2 contains papers 9–13 and
should be attempted after **Practice Papers 1**

Long-style tests, good practice for GL.	9	9	60	30:00
	10	18	60	22:00
Varied-length tests with challenging content. Good practice for pre-test, CEM, GL or any independent school exam.	11	27	30	11:00
	12	35	10	5:00
	13	37	60	22:00

Speed	Score	Time	Notes
Slow	/ 20	:	
Slow	/ 20	:	
Average	/ 20	:	
Average	/ 20	:	
Average	/ 20	:	
Average	/ 20	:	
Fast	/ 30	:	
Fast	/ 10	:	
Average	/ 60	:	
Fast	/ 60	:	
Fast	/ 30	:	
Fast	/ 10	:	
Fast	/ 60	:	

How to use this book

Introduction

These *Practice Papers* have been written to provide final preparation for your 11+ Non-Verbal Reasoning test.

Practice Papers 1 includes eight model papers with a total of 160 questions. There are ...

- four training tests, which include some simpler questions and slower timings designed to develop confidence
- four tests in the style of Pre-Tests and short-format CEM (Centre for Evaluation and Monitoring)/bespoke tests in terms of difficulty, speed and question variation.

The tests cover the standard Non-Verbal Reasoning questions and spatial reasoning questions (where you are asked to imagine a picture from a different perspective), including 3D shapes.

Practice Papers 2 includes five model papers with a total of 220 questions. These papers include ...

- longer-format tests in the GL (Granada Learning)/bespoke style
- a spatial reasoning test with challenging content, speed and question variation to support all 11+ tests.

Practice Papers 1 will help you ...

- become familiar with the way Pre-Test short-format 11+ test questions are presented
- build your confidence in answering the variety of questions set
- work with increasingly difficult questions
- tackle questions presented in different ways
- build up your speed in answering questions to the timing expected in the 11+ tests.

Pre-Test and the 11+ entrance exams

The Galore Park 11+ series is designed for Pre-Tests and 11+ entrance exams for admission into independent schools. These exams are often the same as those set by local grammar schools too. Non-Verbal Reasoning tests now appear in different formats and lengths and it is likely that if you are applying for more than one school, you will encounter more than one type of test. These include:

- Pre-Tests delivered on-screen
- 11+ entrance exams in different formats from GL (Granada Learning) and CEM (Centre for Evaluation and Monitoring)
- 11+ entrance exams created specifically for particular independent schools.

Tests are designed to vary from year to year. This means it is very difficult to predict the questions and structure that will come up, making the tests harder to revise for.

To give you the best chance of success in these assessments, Galore Park has worked with 11+ tutors, independent school teachers, test writers and specialist authors to create these *Practice Papers*. These books cover the styles of questions and the areas of Non-Verbal Reasoning that typically occur in this wide range of tests.

*Because 11+ tests now aim to include variations in the content and presentation of questions, making them increasingly difficult to revise for, **both** books should be completed as essential preparation for all 11+ Pre-Test and Non-Verbal Reasoning tests.*

For parents

These *Practice Papers* have been written to help both you and your child prepare for Pre-Test and 11+ entrance exams.

For your child to get the maximum benefit from these tests, they should complete them in conditions as close as possible to those they will face in the exams, as described in the 'Working through the book' section below.

Timings get shorter as the book progresses to build up speed and confidence.

Some of these timings are very demanding and reviewing the tests after completing the books (even though your child will have some familiarity with the questions) can be helpful, to demonstrate how their speed has improved through practice.

For teachers and tutors

This book has been written for teachers and tutors working with children preparing for both Pre-Test and 11+ entrance exams. The variations in length, format and range of questions are intended to prepare children for the increasingly unpredictable tests encountered, with a range of difficulty developed to prepare them for the most challenging paper and on-screen adaptable tests.

Working through the book

The **Contents and progress record** helps you to understand the purpose of each test and track your progress. Always read the notes in this record before beginning a test as this will give you an idea of how challenging the test will be!

You may find some of the questions hard, but don't worry. These tests are designed to build up your skills and speed. Agree with your parents on a good time to take the test and set a timer going. Prepare for each test as if you are actually going to sit your 11+ (see 'Test day tips' on page 8).

- Complete the test with a timer, in a quiet room. Note down how long it takes you and write your answers in pencil. Even though timings are given, you should complete ALL the questions.
- Mark the test using the answers at the back of the book.
- Go through the test again with a friend or parent and talk about the difficult questions.
- Have another go at the questions you found difficult and read the answers carefully to find out what to look for next time.

The **Answers** are designed to be cut out so that you can mark your papers easily. Do not look at the answers until you have attempted a whole paper. Each answer has a full explanation so you can understand why you might have answered incorrectly.

When you have finished a test, turn back to the 'Contents and progress record' and fill in the boxes:

- Write your total number of marks in the 'Score' box
- Note the time you took to complete ALL the questions in the 'Time' box.

After completing both books you may want to go back to the earlier papers and have another go to see how much you have improved!

Continue your learning journey

When you've completed these *Practice Papers*, you can carry on your learning right up until exam day with the following resources.

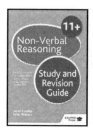 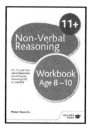 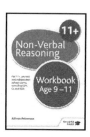 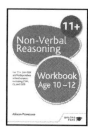   

The *Revision Guide* (referenced in the answers to this book) reviews basic skills in all areas of nonverbal reasoning, and guidance is provided on how to improve in this subject.

Practice Papers 2 contains a further five model papers and answers to increase your familiarity with taking actual tests and to improve your accuracy, speed and ability to deal with a wide range of questions under pressure.

CEM 11+ Non-Verbal Reasoning & Spatial Reasoning Practice Papers contains three practice papers designed for preparation for the CEM-style tests. Each paper is split into short tests in non-verbal reasoning and spatial reasoning. The tests vary in length and format and are excellent for short bursts of timed practice.

GL 11+ Non-Verbal Reasoning Practice Papers contains three practice papers designed for preparation for the GL-style tests.

The *Workbooks* will develop your skills, with over 160 questions to practise in each book. To prepare you for the exam, these books include even more question variations that you may encounter – the more you do, the better equipped for the exams you'll be.

- *Age 8–10*: Increase your familiarity with variations in the question types.
- *Age 9–11*: Experiment with further techniques to improve your accuracy.
- *Age 10–12*: Develop fast response times through consistent practice.

● Paper 1

Test time: 13:30

Look at these sets of pictures. Identify the one that is most unlike the others. Circle the letter beneath the correct answer. For example:

a   b   ⓒ   d   e

1

a   b   c   d   e

2

a   b   c   d   e

3

a   b   c   d   e

4

a   b   c   d   e

5

a   b   c   d   e

Turn over to the next page.

The square given at the beginning is folded in the way indicated by the arrows, and then holes are punched where shown on the final diagram. Identify the answer option that shows what the square would look like when it is unfolded. Circle the letter beneath the correct answer. For example:

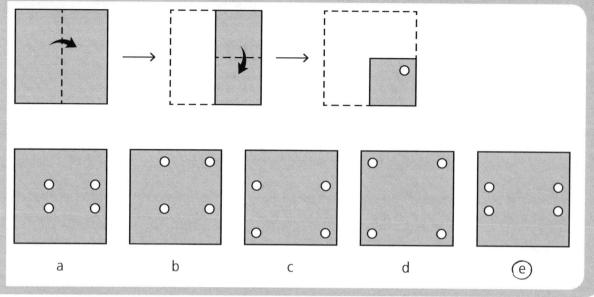

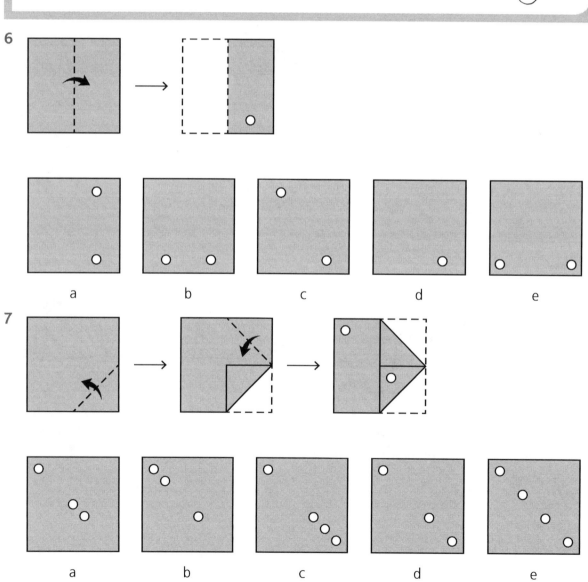

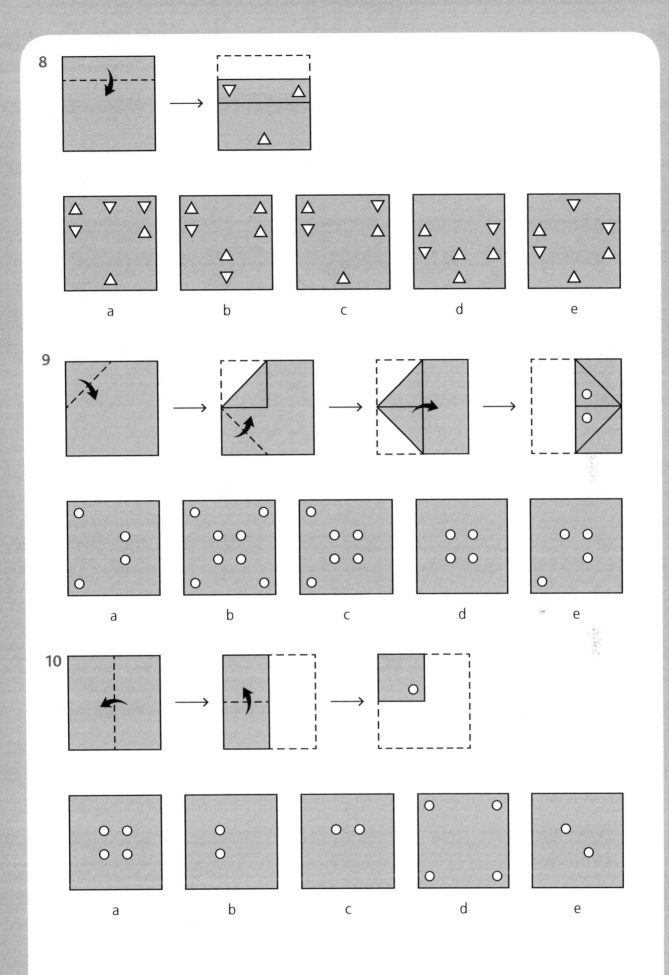

8

9

10

Turn over to the next page.

Look at the two pictures on the left connected by an arrow. Decide how the first picture has been changed to create the second. Now apply the same rule to the third picture and circle the letter beneath the correct answer. For example:

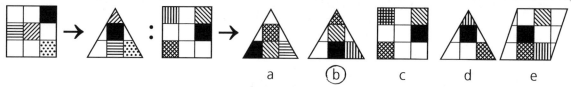

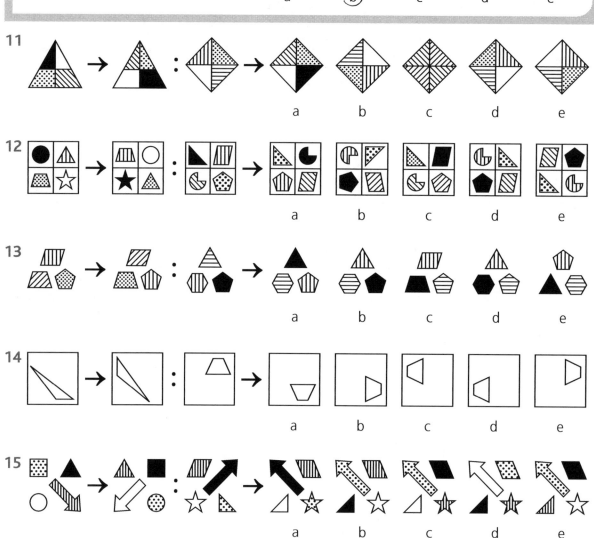

One of the options on the right completes the pattern in the grid on the left. Circle the letter beneath the correct answer. For example:

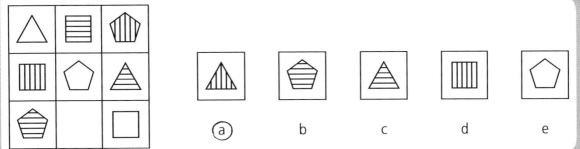

16

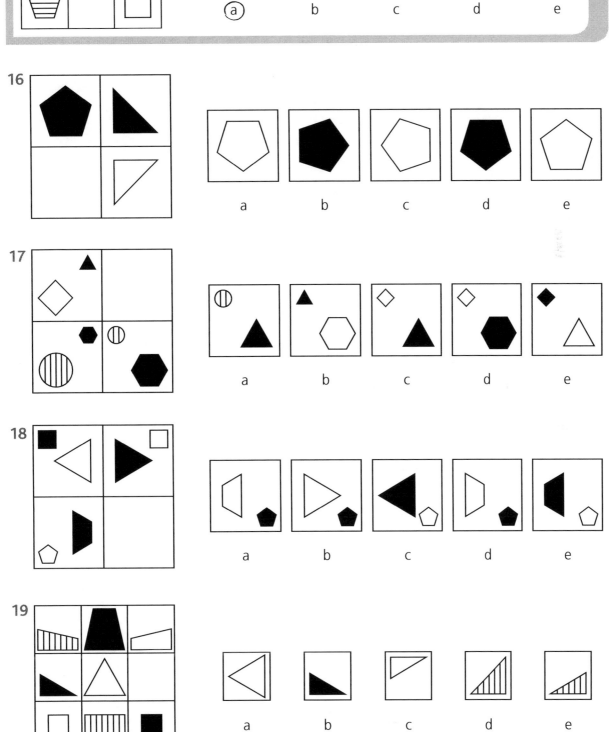

Turn over to the next page.

20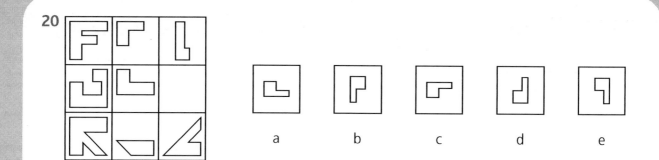

a   b   c   d   e

Paper 2

Look at the first three pictures and decide what they have in common. Then select the option from the five on the right that belongs to the same set. Circle the letter beneath the correct answer. For example:

a   (b)   c   d   e

1

a   b   c   d   e

2

a   b   c   d   e

3

a   b   c   d   e

4

a   b   c   d   e

5

a   b   c   d   e

Turn over to the next page.

15

The small shape on the left can be found in one of the pictures on the right. It might be made up of one or more pieces. The shapes are **not** rotated. Circle the letter beneath the correct answer. For example:

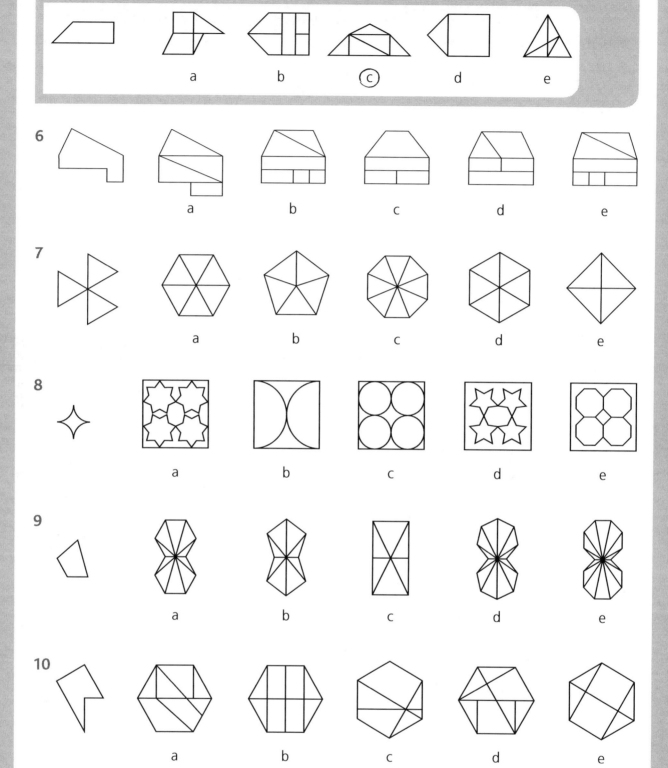

a   b   ⓒ   d   e

6   a   b   c   d   e

7   a   b   c   d   e

8   a   b   c   d   e

9   a   b   c   d   e

10   a   b   c   d   e

The two letters in the small boxes at the right of each large box represent a feature of the shapes in the box. Work out which feature is represented by each letter and apply the code to the box with the dashed lines. Circle the letter beneath the correct answer code. For example:

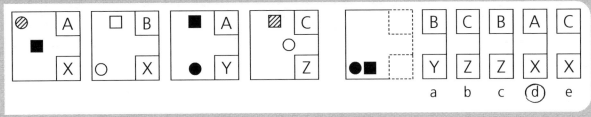

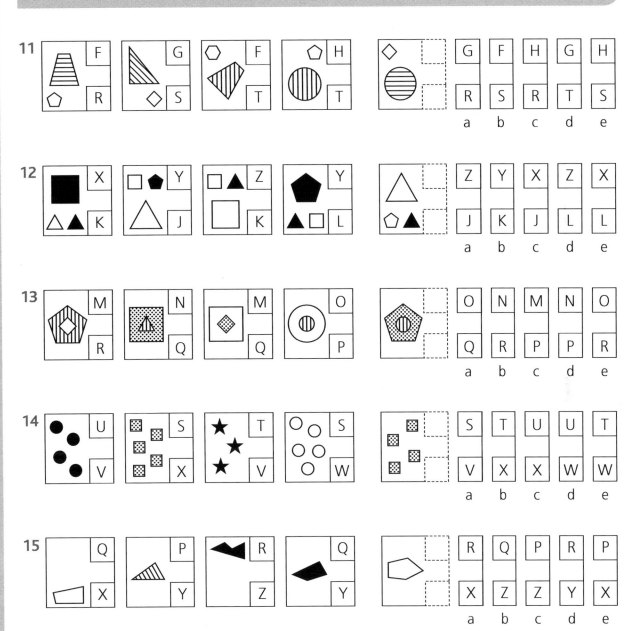

Turn over to the next page.

The five boxes on the left show a pattern that is arranged in a sequence. Choose the answer option that completes the sequence when inserted in the blank box. Circle the letter beneath the correct answer. For example:

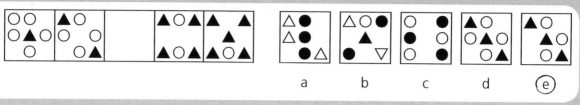

a　　b　　c　　d　　(e)

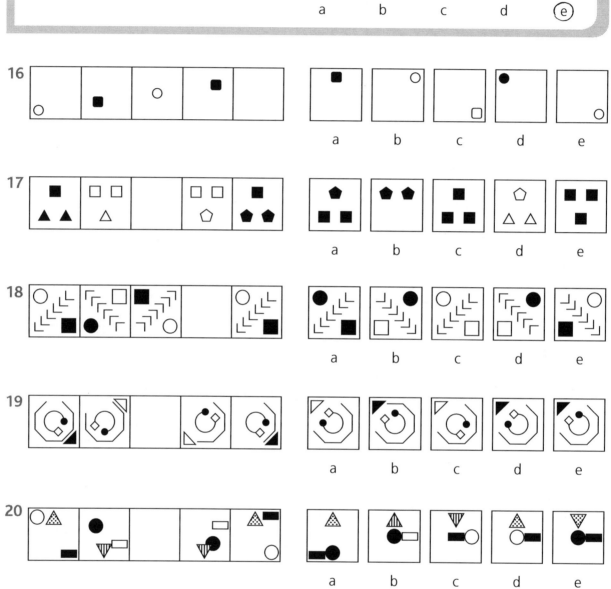

16

a　　b　　c　　d　　e

17

a　　b　　c　　d　　e

18

a　　b　　c　　d　　e

19

a　　b　　c　　d　　e

20

a　　b　　c　　d　　e

Record your results and move on to the next paper.

Score ☐ / 20　Time ☐ : ☐

Paper 3

Test time: 10:00

Look at the first two pictures and decide what they have in common. Then select one of the options from the five on the right that belongs in the same set. Circle the letter beneath the correct answer. For example:

a   b   c   (d)   e

1

    a   b   c   d   e

2

    a   b   c   d   e

3

    a   b   c   d   e

4

    a   b   c   d   e

5

    a   b   c   d   e

Turn over to the next page.

Look at the two pictures on the left connected by an arrow. Decide how the first picture has been changed to create the second. Now apply the same rule to the third picture and circle the letter beneath the correct answer. For example:

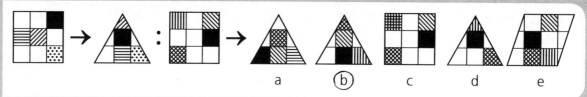

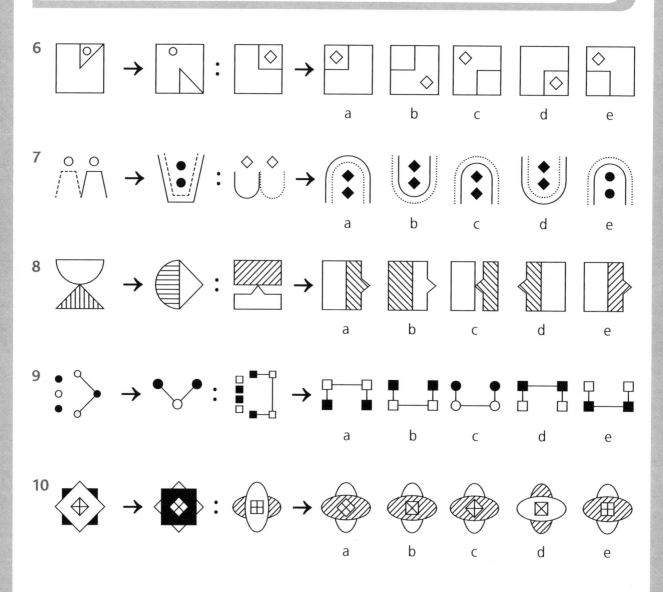

The boxes on the left, or above, show a pattern that is arranged in a sequence. Choose the answer option that completes the sequence when inserted in the blank box. Circle the letter beneath the correct answer. For example:

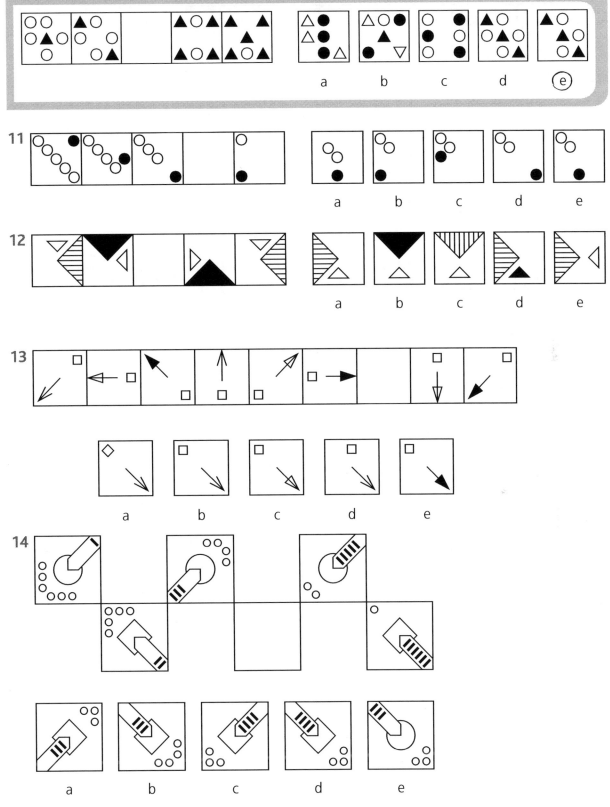

Turn over to the next page.

15

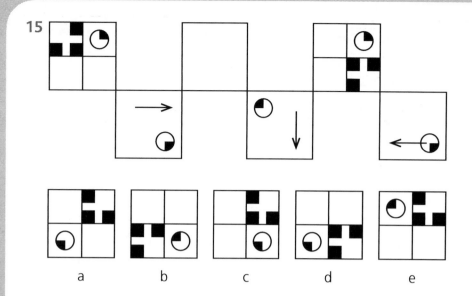

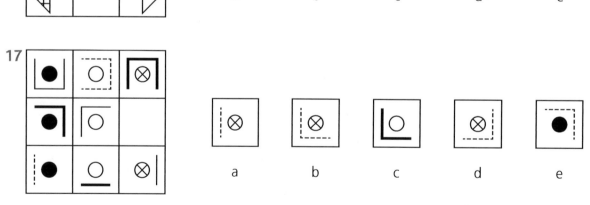

One of the options on the right completes the pattern in the grid on the left. Circle the letter beneath the correct answer. For example:

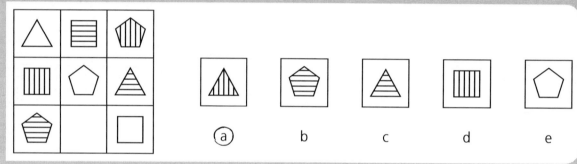

16

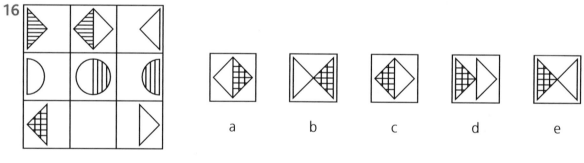

17

18

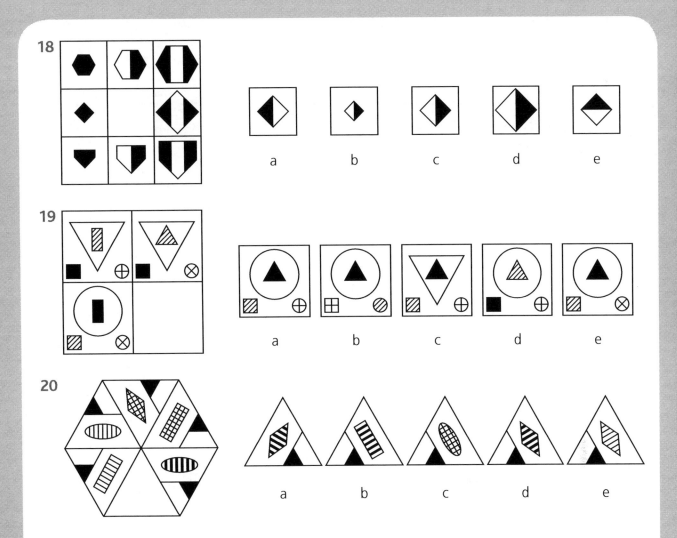

19

20

**Record your results and move
on to the next paper.**

Paper 4

Find the face that would appear **opposite** the face given on the left when the net is folded into a cube. Circle the letter beneath the correct answer. For example:

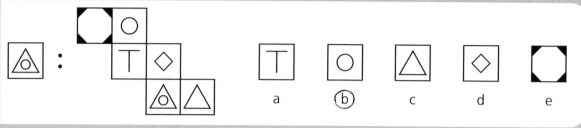

1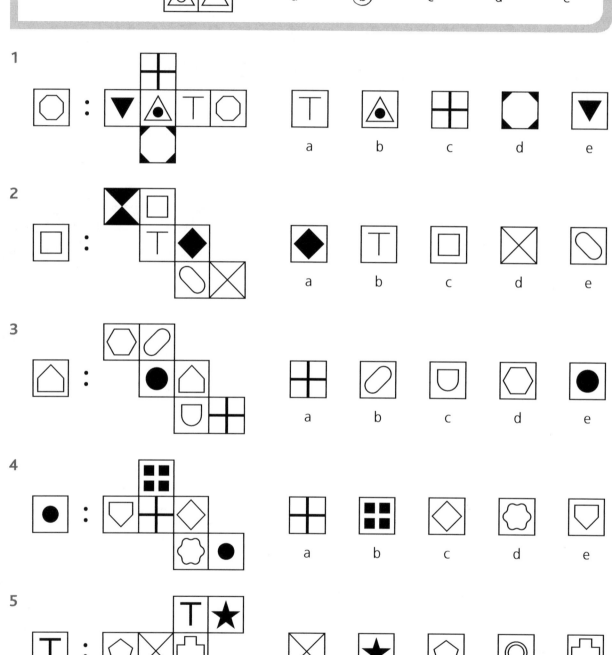

2

3

4

5

Which of the answer options is a 2D plan of the 3D picture on the left, when viewed from above? Circle the letter beneath the correct 2D plan. For example:

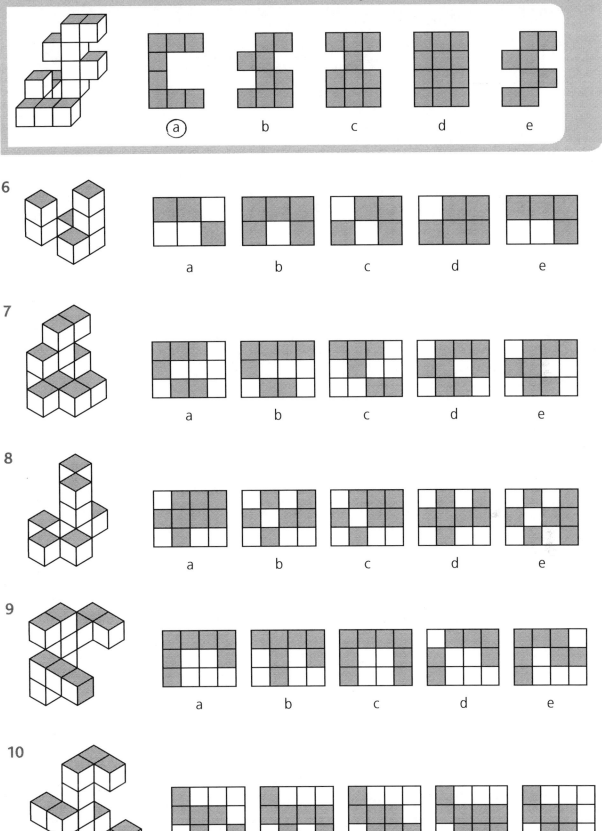

Turn over to the next page.

The picture on the left is rotated to give one of the pictures on the right. Which answer option shows the picture after the rotation? Circle the letter beneath the correct answer. For example:

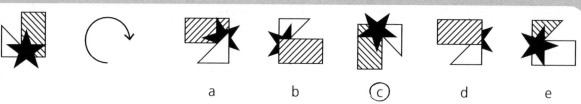

a   b   ⓒ   d   e

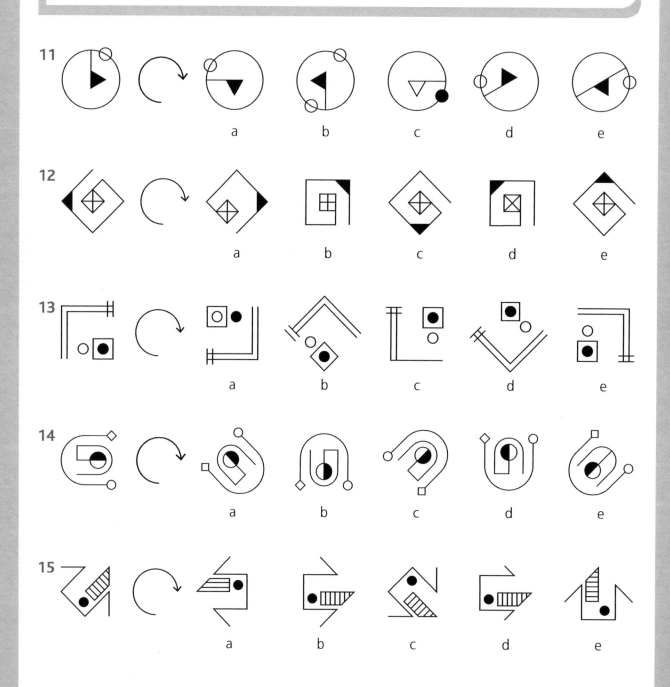

One group of separate blocks has been joined together to make the pattern of blocks shown on the left. Some of the blocks may have been rotated. Circle the letter beneath the blocks that make up the pattern. For example:

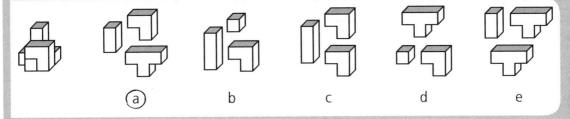

a   b   c   d   e

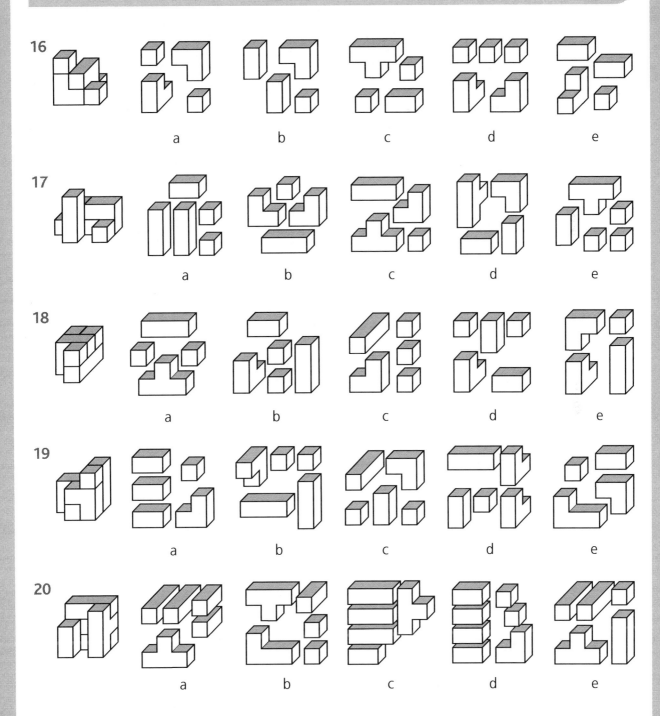

16
a   b   c   d   e

17
a   b   c   d   e

18
a   b   c   d   e

19
a   b   c   d   e

20
a   b   c   d   e

Record your results and move on to the next paper.

Score [ ] / 20   Time [ ] : [ ]

Look at the two sets of shapes on the left connected by arrows and decide how the first shapes have been changed to create the second. Now apply the same rule to the next shape and circle the letter beneath the correct answer from the five options. For example:

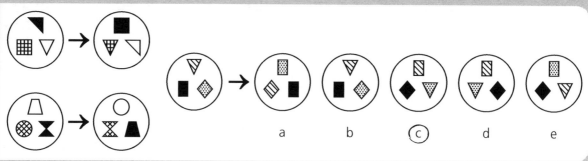

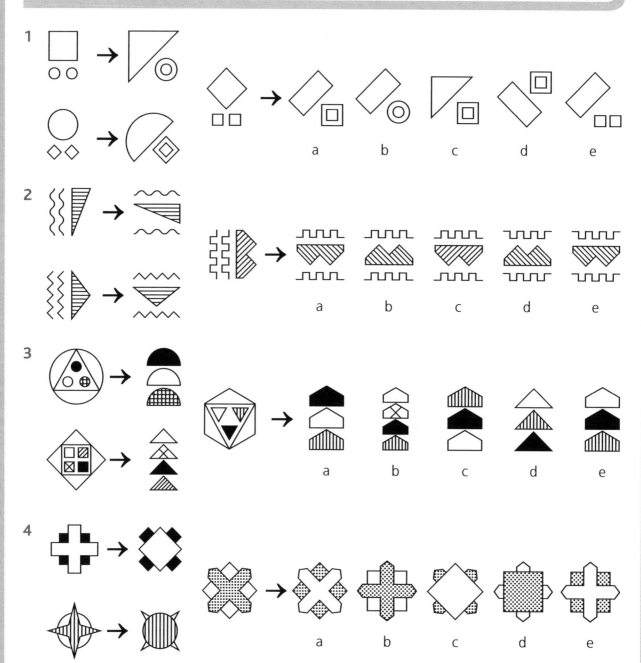

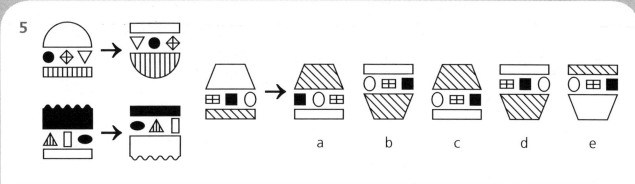

5

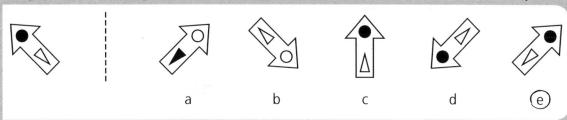

The picture on the left is reflected in a vertical mirror line and is represented by one of the pictures on the right. Circle the letter beneath the correct answer. For example:

a   b   c   d   (e)

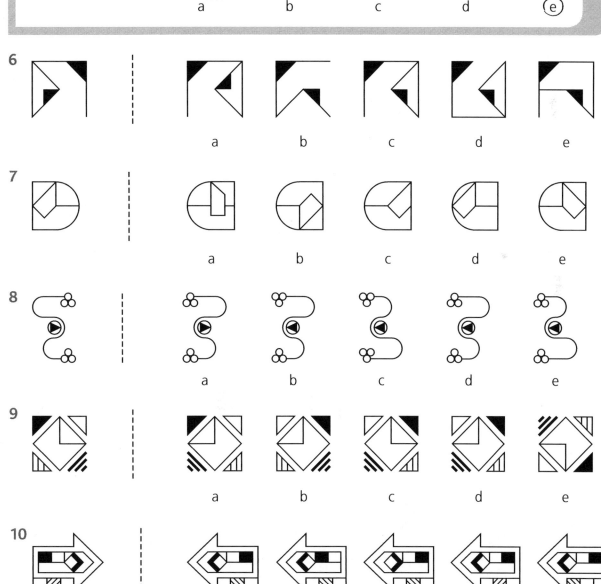

6

a   b   c   d   e

7

a   b   c   d   e

8

a   b   c   d   e

9

a   b   c   d   e

10

a   b   c   d   e

Turn over to the next page.

Each letter represents an individual feature in the picture next to it. Work out which feature is represented by each letter. Apply the code to the picture in the box and circle the letter beneath the correct answer code. For example:

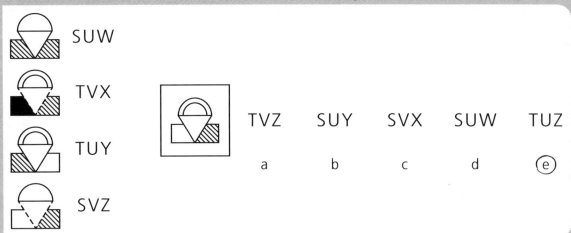

SUW

TVX

TUY

SVZ

TVZ	SUY	SVX	SUW	TUZ
a	b	c	d	(e)

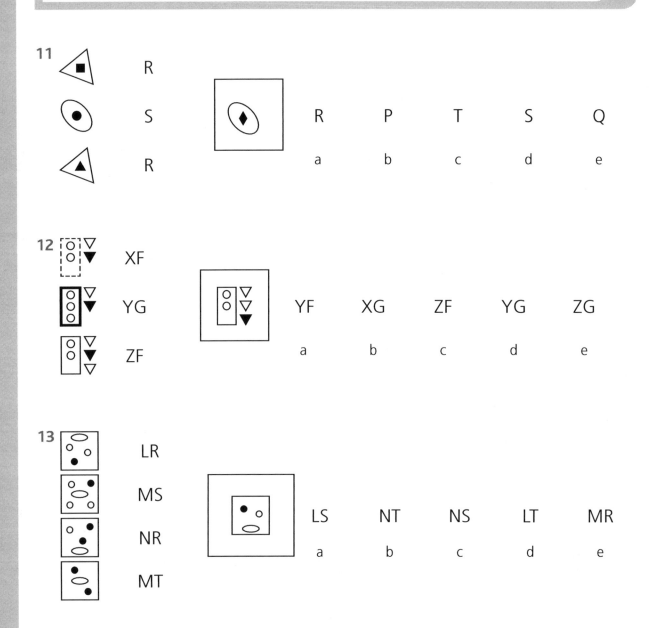

11

R

S

R

R	P	T	S	Q
a	b	c	d	e

12

XF

YG

ZF

YF	XG	ZF	YG	ZG
a	b	c	d	e

13

LR

MS

NR

MT

LS	NT	NS	LT	MR
a	b	c	d	e

30

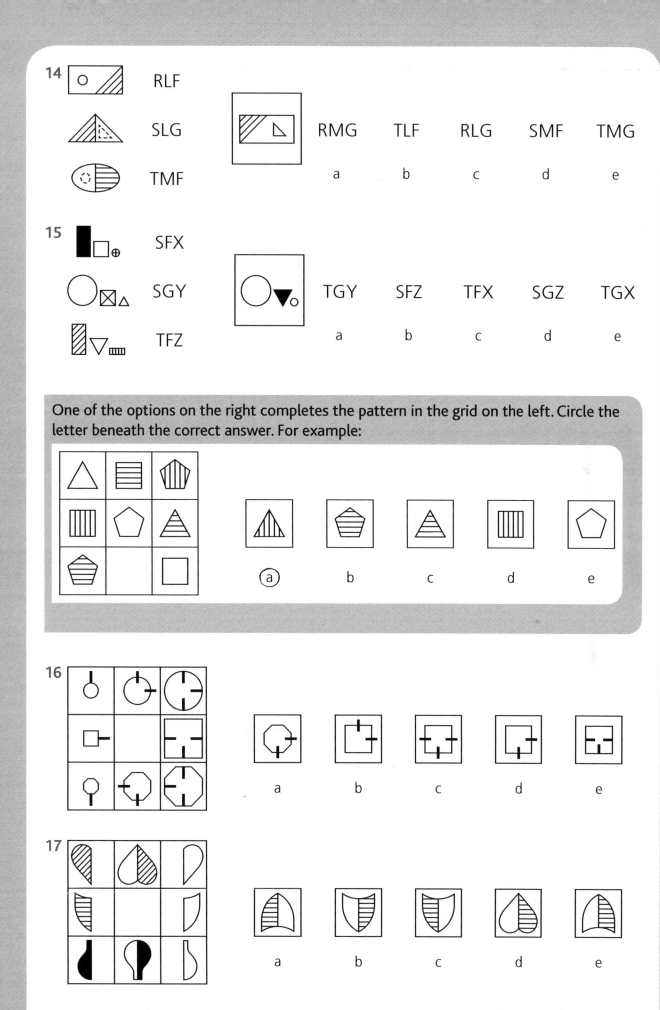

14

○ ▱ RLF

◭ SLG

⬭ TMF

▱ ◿   RMG   TLF   RLG   SMF   TMG

    a   b   c   d   e

15

▮ ▫ ⊕ SFX

◯ ⊠ △ SGY

◪ ▽ ▭ TFZ

◯ ▼ ◦   TGY   SFZ   TFX   SGZ   TGX

    a   b   c   d   e

One of the options on the right completes the pattern in the grid on the left. Circle the letter beneath the correct answer. For example:

    (a)   b   c   d   e

16

    a   b   c   d   e

17

    a   b   c   d   e

Turn over to the next page.

18

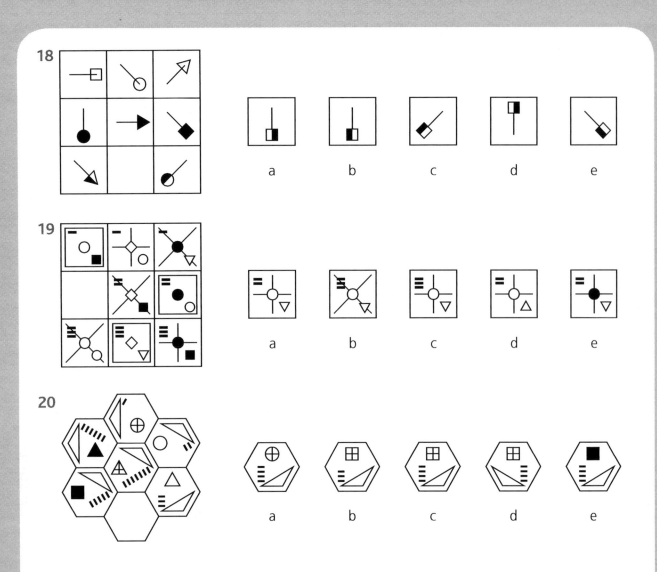

a　　b　　c　　d　　e

19

a　　b　　c　　d　　e

20

a　　b　　c　　d　　e

Record your results and move on to the next paper.

Score ☐ / 20　Time ☐ : ☐

Paper 6

Find the cube that can be made from the net shown on the left. Circle the letter beneath the correct answer. For example:

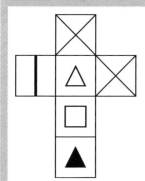

a   b   c   (d)   e

1

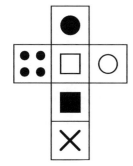

a   b   c   d   e

2

a   b   c   d   e

3

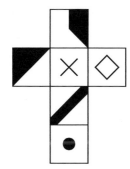

a   b   c   d   e

Turn over to the next page.

4

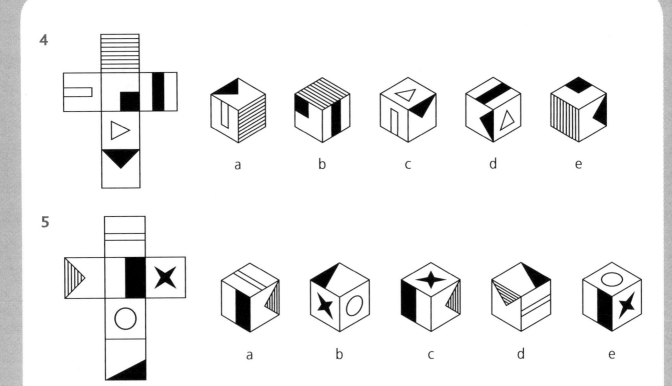

5

In questions 6–10 you will see a rotated version of one of the 3D diagrams shown (a, b, c, d, e). Circle the letter of the answer option that indicates the matching shape above.

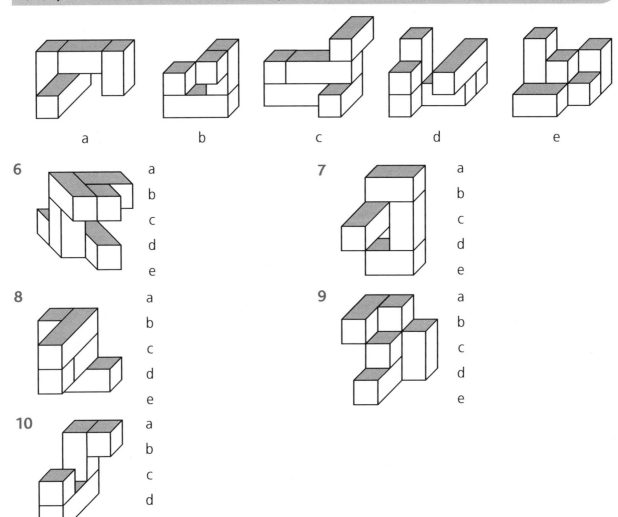

a   b   c   d   e

6
a
b
c
d
e

7
a
b
c
d
e

8
a
b
c
d
e

9
a
b
c
d
e

10
a
b
c
d
e

The pattern of circles in the left-hand box has been hidden within the pattern of circles in the right-hand box. Carefully draw a line around the pattern once you find it. For example:

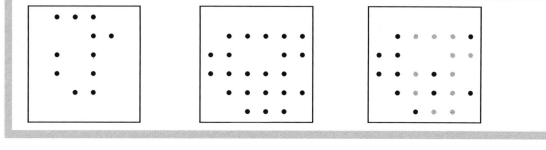

11

12

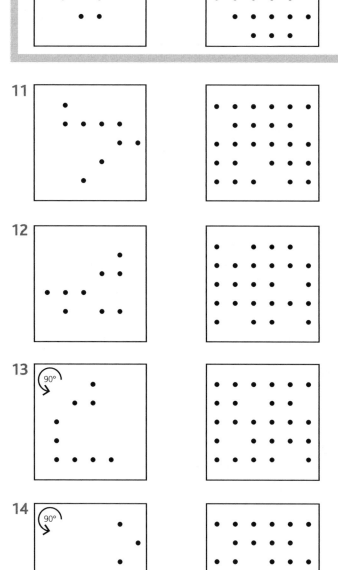

13 90°

14 90°

15 90°

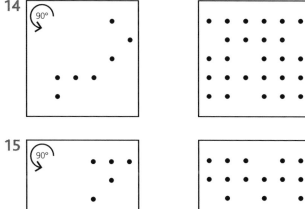

Turn over to the next page.

One group of separate blocks has been joined together to make the pattern of blocks shown on the far left. Some of the blocks may have been rotated. Circle the letter beneath the blocks that make up the pattern. For example:

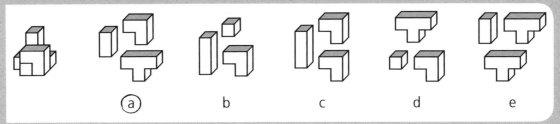

a   b   c   d   e

16
a   b   c   d   e

17
a   b   c   d   e

18
a   b   c   d   e

19
a   b   c   d   e

20
a   b   c   d   e

Record your results and move on to the next paper.

Score [ ] / 20   Time [ ] : [ ]

36

Paper 7

Test time: 11:00

Look at these sets of pictures. Identify the one that is most unlike the others. Circle the letter beneath the correct answer. For Example:

a   b   ⓒ   d   e

1

a   b   c   d   e

2

a   b   c   d   e

3

a   b   c   d   e

4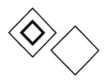

a   b   c   d   e

5

a   b   c   d   e

Turn over to the next page.

Look at the two sets of shapes on the left connected by arrows and decide how the first shapes have been changed to create the second. Now apply the same rule to the next shape and circle the letter beneath the correct answer from the five options. For example:

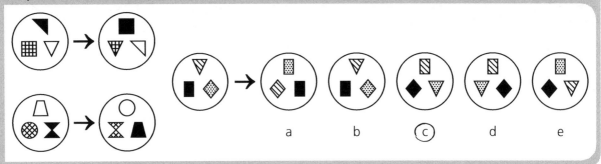

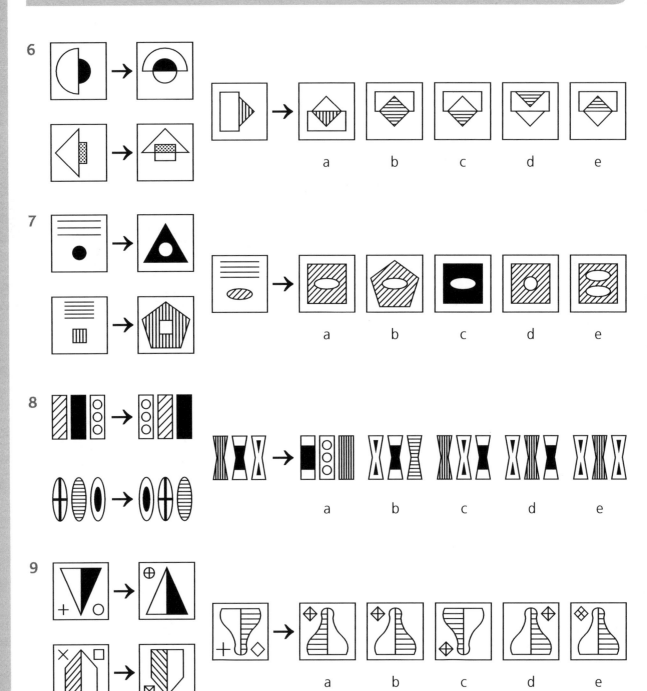

10

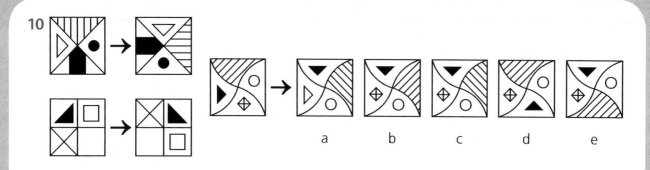

Identify the diagram that shows how the plan on the left will appear when it is folded in along the dashed lines. Circle the letter beneath the correct answer. For example:

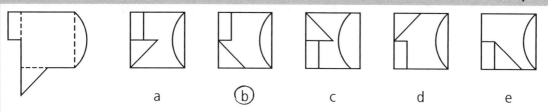

a   (b)   c   d   e

11

12

13

14

15

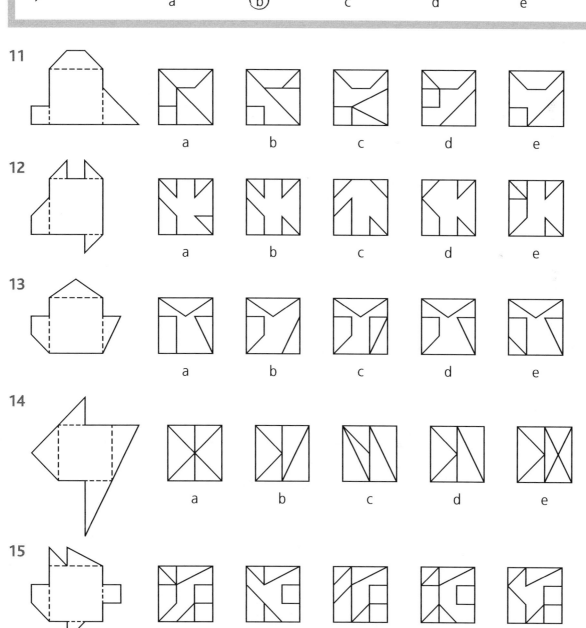

Turn over to the next page.

The picture on the left is reflected in a vertical mirror line and is represented by one of the pictures on the right. Circle the letter beneath the correct answer. For example:

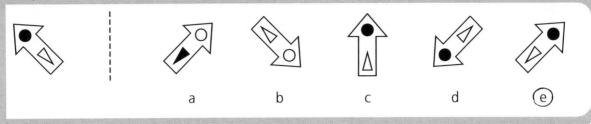

a   b   c   d   (e)

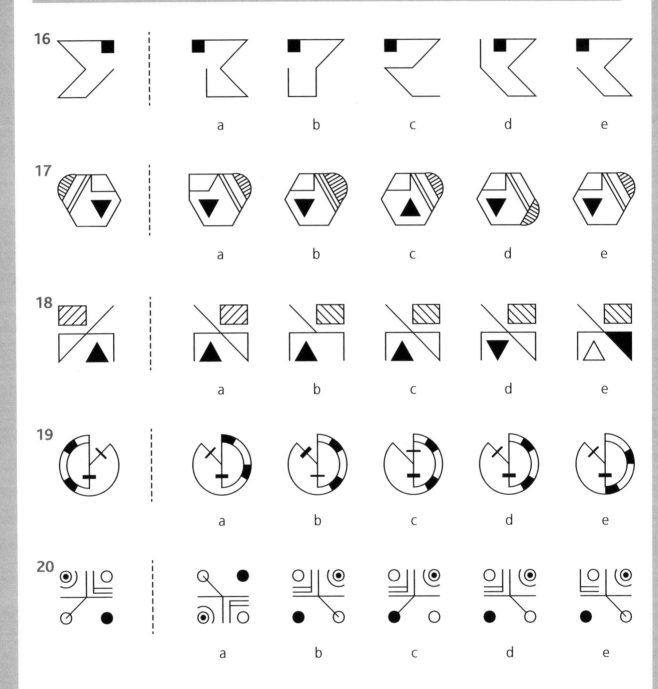

16

a   b   c   d   e

17

a   b   c   d   e

18

a   b   c   d   e

19

a   b   c   d   e

20

a   b   c   d   e

Each letter represents an individual feature in the picture next to it. Work out which feature is represented by each letter. Apply the code to the picture in the box and circle the letter beneath the correct answer code. For example:

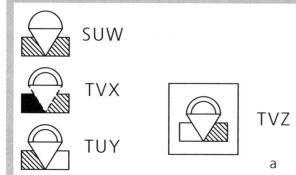

SUW

TVX

TUY

SVZ

TVZ   SUY   SVX   SUW   TUZ

a   b   c   d   (e)

21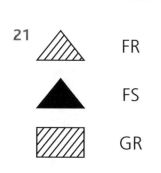

FR

FS

GR

FR   GR   GS   FS   GT

a   b   c   d   e

22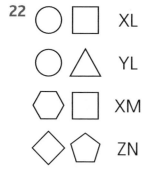

XL

YL

XM

ZN

XN   YN   ZL   YM   ZM

a   b   c   d   e

23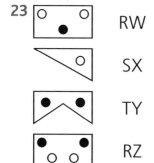

RW

SX

TY

RZ

TW   RX   SY   RY   TX

a   b   c   d   e

Turn over to the next page.

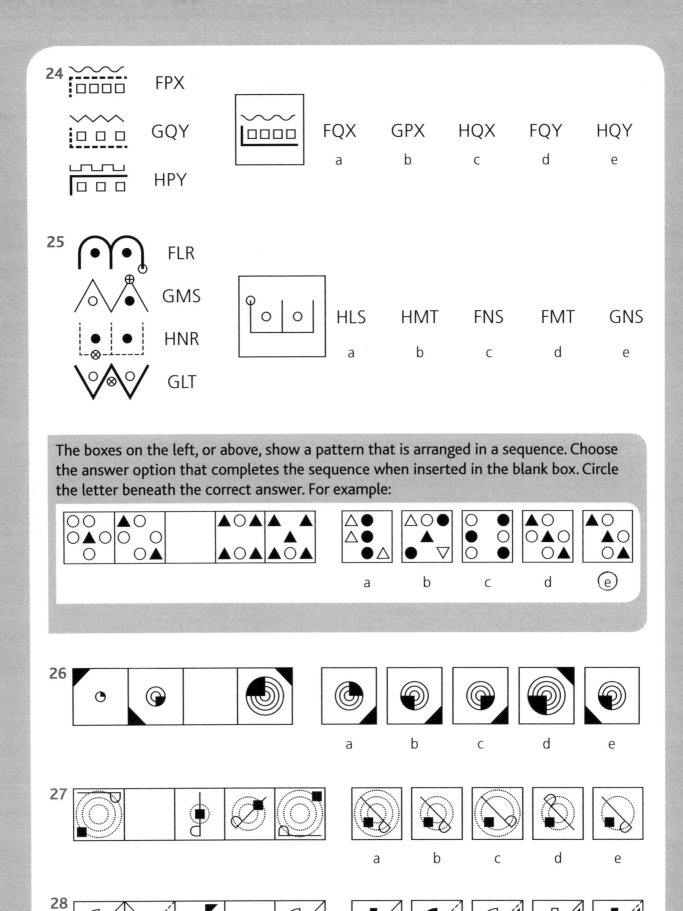

24

FPX

GQY

HPY

FQX	GPX	HQX	FQY	HQY
a	b	c	d	e

25

FLR

GMS

HNR

GLT

HLS	HMT	FNS	FMT	GNS
a	b	c	d	e

The boxes on the left, or above, show a pattern that is arranged in a sequence. Choose the answer option that completes the sequence when inserted in the blank box. Circle the letter beneath the correct answer. For example:

a b c d (e)

26

a b c d e

27

a b c d e

28

a b c d e

29

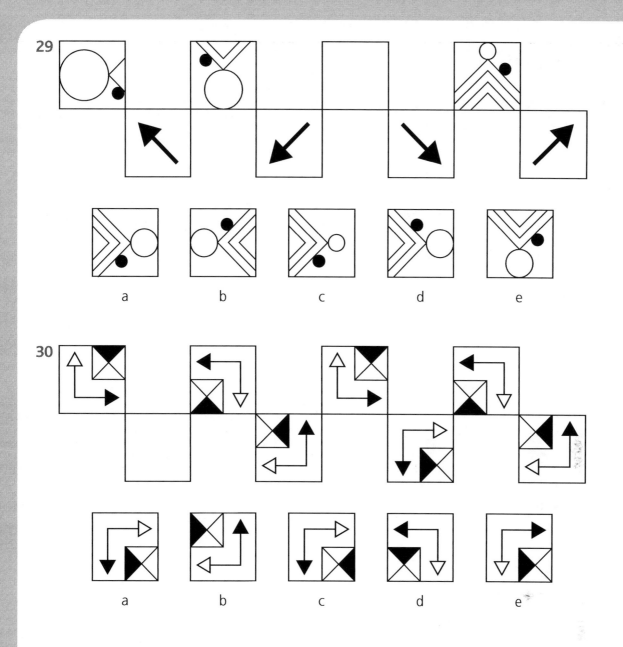

30

Record your results and move on to the next paper.

Score ☐ / 30　Time ☐ : ☐

Paper 8

Find the cube, or other 3D shape, that can be made from the net shown on the left. Circle the letter beneath the correct answer. For example:

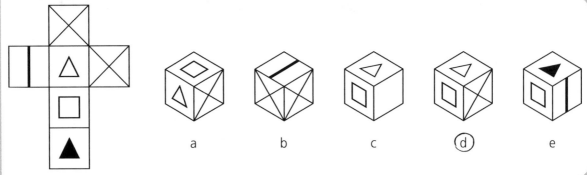

1

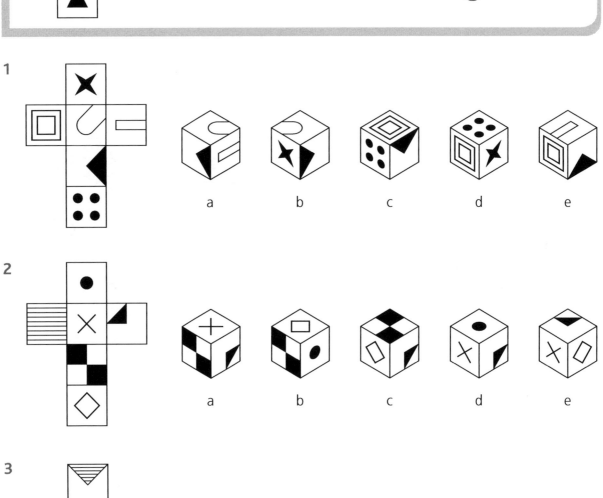

2

3

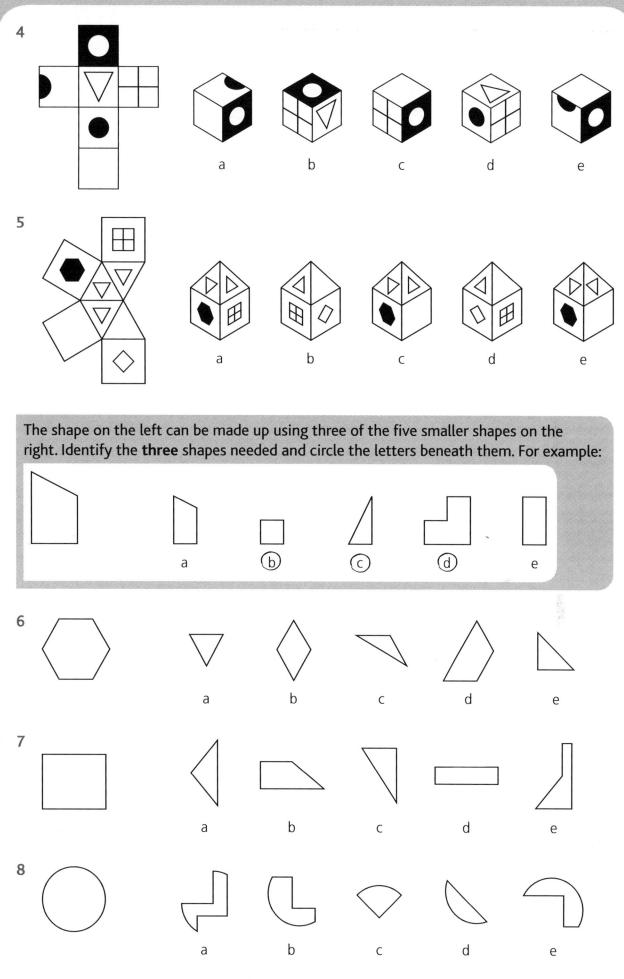

4

5

The shape on the left can be made up using three of the five smaller shapes on the right. Identify the **three** shapes needed and circle the letters beneath them. For example:

a   (b)   (c)   (d)   e

6

a   b   c   d   e

7

a   b   c   d   e

8

a   b   c   d   e

Turn over to the next page.

9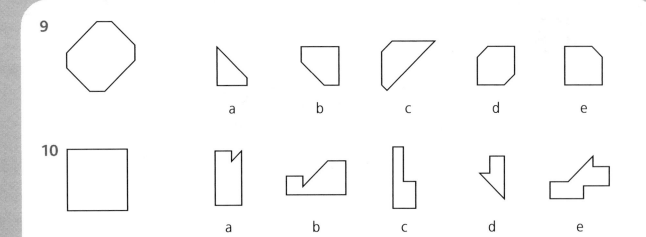

a   b   c   d   e

10

a   b   c   d   e

Answers

Please note all questions are worth one mark.

Paper 1 (page 9)

1 **d** **shading** – four of the shapes are shaded white, only shape **d** has a different shading.
Distractors: **shape** – the shapes are unimportant.

2 **a** **number** – four of the shapes have five lines, shape **a** has four lines.
Distractors: **position** – whether there is a crossover is unimportant; **size** – the length of each line is unimportant.

3 **c** **position/orientation** – four of the arrows are pointing to the right, only **c** is pointing to the left.
Distractors: **shape** – the arrowhead is unimportant; **size** – the length of the arrows is unimportant; **style** – the thickness of the arrows is unimportant.

4 **d** **shading** – all of the squares have diagonally striped shading, but in **d** it is going from bottom-left to top-right.
Distractors: **shape** – (a) the pentagon is unimportant; (b) the circle is unimportant; **position/orientation** – the direction of the arrow is unimportant.

5 **e** **shape** – the black shape is the same as the overlapping area except in e; **rotation** – the black shape has been rotated.
Distractors: **rotation** – the size of the rotation is unimportant; **shape** – the type of white shapes used are unimportant.

6 **b** The fold acts like a vertical line of reflection. The hole is punched centrally at the base so will be in a central position on each side of the fold when the paper is opened.

7 **d** The first fold acts like a diagonal line of reflection while the second fold is a distraction, as the hole on the left is not punched through the folded area. The hole in the bottom right will be copied further into the bottom-right corner, close to the fold line when the bottom fold is opened out.

8 **c** The fold acts like a horizontal line of reflection. The top two holes are punched close to the fold so the new holes will be similarly close to the other side of the fold when the paper is opened, but will be reflected and pointing in the opposite direction. The hole at the bottom is not in the folded area so will not be copied.

9 **c** The first two folds act as diagonal lines of reflection while the final fold acts as a vertical line of reflection. Both holes will go through three layers, so there will be six holes. The bottom layer is the rectangle shape so there will be two holes in the position shown on the final diagram, then two further holes to the left of these when the vertical fold is opened. The last two holes will appear in the top and bottom corners on the left when the two diagonal folds are opened out.

10 **a** The first fold acts as a vertical line of reflection and the second acts as a horizontal line of reflection. The single hole goes through four layers so there will be four holes. Opening the second fold reveals a hole immediately beneath the punched hole, and opening the first fold reveals two more holes to the right of them.

11 **b** **rotation** – the shading styles move two segments clockwise around the shape.
Distractor: **shading** – the example uses similar styles of shading so difficult to spot the correct answer.

12 **d** **shading** – the four shading styles are retained, but they move one box anticlockwise; **position** – the shapes move one box clockwise.
Distractor: **rotation** – the shapes are not rotated.

13 **d** **shading** – the shading patterns move one shape clockwise.
Distractor: **shape** – the shapes do not change.

14 **e** **reflection** – the box is reflected in a diagonal line from bottom left to top right.

15 **c** **reflection** – the group of shapes is reflected in a vertical line; **shading** – the shading of shapes on the diagonal top left to bottom right swap places.
Distractor: **shading** – the shading on the shapes top right to bottom left stays the same.

16 **a** **reflection** – within each column the shapes are reflected in a horizontal line; **shading** – the top row has black shapes and the bottom row has white shapes.
Distractors: **rotation** – the incorrect answers bring in rotation as a distraction.

17 **c** **size** – the shapes swap size as you move across the row; **translation** – the shapes swap position as you move across the row.
Distractor: **shading** – the shading of the shapes doesn't change.

18 **a** **reflection** – the shapes are reflected in a vertical line as you go across the row; **shading** – (a) the black shape becomes white, (b) the white shape becomes black.

19 **e** **translation/construction** – the shapes in the first and last columns are combined to make the shape in the middle column; **shading** – each row and column has one white shape, one black shape and one stripy shape; **position** – the shapes in the first column will be rotated anticlockwise to make part of the middle shape, while the shapes in the last column will be rotated clockwise to make part of the middle shape.
Distractor: **position** – the arrangement of shading is random, with the only rule being that each row and column contains one of each type of shading.

20 **c** **construction** – the shapes in columns two and three combine to make the shape in column one; **translation** – the shapes in column two have been moved vertically to the opposite end of the box from where they are in the larger shape; **rotation** – the shapes in the third column have been rotated 90° anticlockwise compared to the first column.
Distractor: **reflection** – some of the possible answers have been reflected.

Paper 2 (page 15)

1 **a** **shape** – the shapes are all right-angled triangles.
Distractors: **shading** – the shading of the shapes is unimportant; **position** – the orientation of the shapes is a distraction designed to mask the right angles.

2 **e** **shape** – the shapes are all quadrilaterals.
Distractor: **position** – the orientation of the shapes is unimportant.

3 **d** **reflection** – the shapes are reflected in a vertical line; **position** – the shapes are touching along the mirror line.
Distractor: **shading** – the shading is a distraction.

4 **e** **shading** – the rectangle is always spotty; **position** – the rectangle is always outside the star and the circle is inside the star.
Distractors: **shading** – how the star is shaded; **proportion** – the size of the shapes.

5 **b** **number** – there are always six small shapes; **shading** – two of the small shapes are always black, the other four are always white.
Distractors: **position** – (a) the position of the larger white circle is unimportant, (b) the position of all the smaller shapes is unimportant; **number** – the number in each group of small shapes is unimportant.

6 **b** The most identifiable feature of the small shape is the angle at the top. Options **a**, **b** and **d** have this angle, the others can be excluded. The bottom of the small shape is a small square on the right, with a similar height rectangle above it. Option **a** is too tall, **d** has a rectangle and no square, leaving option **b**.

7 **d** The small shape is made up of a number of triangles with the far left triangle having a vertical edge. Options **c** and **d** have triangles with vertical edges. Option **c** has smaller triangles, leaving option **d**.

8 **c** The small shape is made up of four curved lines. Options **a**, **d** and **e** only use straight lines so can be excluded. Option **b** only has two curved lines that meet, leaving option **c**.

9 **a** The small shape is a quadrilateral with a horizontal base with two lines going up and to the left from the base. Option **c** does not have two lines

going up and to the left from any horizontal line. Options **b**, **d** and **e** do have the lines, but the base is not long enough, leaving option **a**.

10 **d** The most identifiable feature of the small shape is a right angle with a line descending from it half way down the shape. The small shape also has some angled lines. Only options **b** and **d** have the right angle with the descending line, but **b** does not have the angled lines, leaving option **d**.

11 **c** **shape** – the first letter represents the main shape, F is for a quadrilateral, G is for a triangle, H is for a circle; **shading** – the second letter describes the shading of the shape, R is for a horizontal pattern, S for a diagonal pattern and T for a vertical pattern.
Distractor: **shape** – the smaller shape is unimportant.

12 **a** **number** – the first letter represents the number of sides on the shapes inside the box, X is for 10 corners, Y is for 12 corners and Z is for 11 corners; **shape** – the second letter represents the largest shape, J is for a triangle, K is for a square and L is for a pentagon.
Distractors: **shading** – the shading of the shapes is unimportant; **position** – the position of the shapes within the box is unimportant.

13 **e** **shape** (a) the first letter represents the smaller shape, M is for a square, N is for a triangle and O is for a circle, (b) the second letter represents the larger shape, P is for a circle, Q is for a square and R is for a pentagon.
Distractor: **shading** – the shading of the shapes is unimportant.

14 **c** **number** – the first letter represents the number of shapes, S is for 5 shapes, T is for 3 shapes and U is for 4 shapes; **shading** – the second letter represents the shading, V is for black, W is for white and X is for spotty.
Distractors: **shape** – the shapes are unimportant.

15 **d** **number** – the first letter represents the number of sides on the shapes, P is for 3, Q is for 4 and R is for 5; **position** – the second letter represents the position of the shape within the box, X is for bottom, Y is for middle, Z is for top.
Distractor: **shading** – the shading of the shapes is unimportant.

16 **b** **shape** – the shapes alternate between a circle and a rounded square; **shading** – the circles are white and the rounded squares are black; **translation** – the shapes slowly move from the bottom-left corner to the top-right corner.

11+ Non-Verbal Reasoning Practice Papers 1 published by Galore Park

17 c **number** – (a) there are always three shapes inside the box, (b) the total number of sides for the shapes increased by one each time; **shading** – the shapes in the boxes alternate between black and white shading; **position** – the shapes are arranged in a triangle that points up when the shading is black and down when the shading is white.

18 b **rotation** – (a) the arrowheads rotate clockwise around the box, (b) the square and circle move anticlockwise around the box; **shading** – the square and circle alternate the black and white shading.

19 d **rotation** – (a) the central shape rotates 90° clockwise each time, (b) the triangle rotates 90° anticlockwise as it moves anticlockwise around the box; **shape** – the octagon loses one side each time; **shading** – the triangle alternates between black and white.
Distractor: **position/shading** – the black circle and white square do not change their shading or position.

20 d **translation** – (a) the circle moves from the top left to the bottom right in small steps, (b) the rectangle moves from the bottom right to the top right in small steps; **shading** – (a) the circle and rectangle alternate between black and white shading, (b) the triangle alternates between spotted and striped shading; **reflection** – the triangle reflects around a central horizontal line.

Paper 3 (page 19)

1 b **shape** – Each picture is made of an oval and diamond shape; **shading** – the right-hand side of the diamond is shaded (when the shape is rotated so that the oval is at the top).
Distractor: **rotation** – how the shape is rotated is unimportant.

2 c **number** – (a) each picture has six sides, (b) each picture has three small shapes; **shape** – the shapes within each picture are identical.

3 e **line style** – the triangle is surrounded first by a dashed then a solid line, working outwards; **position** – (a) the lines fit around one of the equal angles, not the right angle, (b) the smaller triangle sits on the long side of the larger triangle.
Distractor: **shading** – the shading in the smaller triangle is unimportant.

4 c **size** – there is a large, medium-sized and small shape in each picture; **shape** – (a) the large shape is always a quadrilateral, (b) the medium-sized shape is always a triangle; **position** – the circle is always inside the triangle.
Distractors: **shading** – the shading of the circle is unimportant.

5 d **position** – (a) Each picture has two lines crossing the small square from corner to corner, (b) the white circle is always in a segment next to the black segment, (c) the square is always in a corner.
Distractor: **position** – the corner the square is placed in can be at the top or bottom of the outer box.

6 c **reflection** – (a) the small shape reflects vertically about the centre of the box, and so moves from the top right to the top left of the box, (b) the large shape reflects on a horizontal line to appear at the bottom right of the box.

7 c **reflection** – the picture reflects in a horizontal mirror line; **translation/size** – the shapes change size and the left-hand shape fits inside the right-hand shape; **shape** – the small shapes remain the same (although change shading and move inside the larger shapes).

8 a **rotation** – the two pictures rotate 90° clockwise; **translation** – the two pictures move to join with the bottom picture to the right, **shading** – the shading of the two pictures swaps.

9 e **shading** – the two shapes on the points and corners swap shading; **rotation** – picture rotates 90° clockwise, enlarges.
Distractor: **shape** – the small floating shapes are unimportant.

10 b **position** – the two large shapes swap position; the back comes to the front; rotation – the cross rotates 45°.
Distractor: **rotation** – the small square appears to move because of the position of the cross and changing layers, but stays in the same position.

11 e **number** – the number of white circles decreases, working left to right across the sequence; **position** – the black circle moves half a side clockwise around the box, working left to right across the sequence.
Distractor: **position** – from the first three boxes in the sequence, the black circle visually appears to follow a diagonal pattern.

12 a **rotation** – (a) the large triangle rotates 90° anticlockwise and moves around the edges of the box, (b) the small triangle rotates 90° clockwise, sitting centrally on the side; **shading** – the large triangle alternates in shading from striped to black.

13 b **rotation** – the arrow rotates 45° clockwise; **position** – the square moves half a side clockwise; **shape** – the arrow head changes three times and then repeats the sequence.

14 d **number** – (a) the number of small white circles decreases by one, working up and down across the sequence, (b) the number of black lines increases by one; **rotation** – the picture rotates 90° and moves clockwise around the box by one corner, working up and down across the sequence; **shape** – the shape at the end of the 'arrow' is a circle on the top row and square on the bottom row.

15 a **position** – (a) the circle moves one corner clockwise around the box, moving across the sequence, counting all boxes, (b) the white 'cross' moves one corner clockwise across the top sequence only, (c) the white quarter of the 'cross' shape moves one corner clockwise across the top sequence only; **rotation** – (a) the arrow rotates 90° clockwise as it moves around the box, (b) the shaded quarter-circle rotates 90° clockwise within the circle.

16 **e** **reflection** – (a) the left-hand column reflects in a vertical line on the right, (b) the right-hand column reflects in a vertical line on the left; **shading** – the shading does not change when the shapes are reflected.

17 **b** **shading/pattern** – shading or pattern changes between columns; **number** – the number of lines changes between rows; **line style** – the line style follows a pattern from top left to bottom right; **rotation** – the lines rotate 90° anticlockwise moving across the rows.

18 **c** **size** – the shapes grow moving across the columns; **proportion** – the complete shape is shaded in column 1, half (the right half) in column 2 and two-thirds in column 3; **shape** – the shapes change between rows.

19 **a** **shape** – (a) the two shapes at the base remain unchanged, (b) the large shape changes between rows, (c) the small, inner shape changes between columns; **shading** – the shading of the small square and the inner shape change between rows; **position** – the position of the cross in the circles matches on the diagonals.

20 **d** **rotation** – the black 'flag' rotates 60° clockwise moving round the hexagon; **shape** – the shapes match in opposite segments (they do not rotate); **shading** – the shading works in pairs with the two segments next to each other carrying the same shading (the shading also rotates 60° clockwise).

Paper 4 (page 24)

1 **b** When the net is folded to make the cube, the face with the octagon will be on the opposite face to the triangle with black circle in the centre. Imagine rolling up the central line of squares: the faces with the black triangle and 'T' shape will be next to the triangle with the black circle, so the side underneath must be the octagon.

2 **e** When the net is folded to make the cube, the face with the white square will be on the opposite face to the white lozenge shape. Imagine pinching together the faces with the white and black squares, then folding down the face with the 'T' shape. The face with the lozenge will then fold underneath to be opposite the white square.

3 **d** When the net is folded to make the cube, the face with the house shape will be on the opposite face to the hexagon. Imagine pinching together the faces with the lozenge and house shapes, then folding the face with the black circle downwards. The hexagon will then be free to fold underneath the cube, making it opposite to the house shape.

4 **a** When the net is folded to make the cube, the face with the black circle will be on the opposite face to the face with the black cross. Imagine pinching together the face with the flower shape and the face with the black cross. Next fold down the white diamond to form the side next to the flower and cross. This leaves the black circle free to fold underneath the cube so that it is opposite the black cross.

5 **d** When the net is folded to make the cube, the face with the 'T' will be on the opposite face to the face with the double circle. Imagine pinching together the face with the 'T' and the face with the 'X'. Next fold down the double circle. The 'T' is now opposite the double circle.

6 **c** The far left column is on the front row, so is the bottom-left square on the plan. This touches the cube on the back row at its far right corner. There are two cubes on this back row (so two squares on the top row of the plan), with the single cube at the front of the right-hand column (the square on the bottom row to the right).

7 **e** The bottom layer of the stack of cubes has two cubes at the front, so the bottom of the 2D plan must have two squares in the row at the base. There are then two cubes behind, one above the front cube on the left and one to the left of this cube. This must mean that there is a gap in the plan on the bottom row (left). There is therefore another gap to the right on the bottom row (since there are only two cubes in this row). The final row of cubes has three faces showing at the top, beginning at the same point as the cubes on the base of the plan, so our plan is complete.

8 **b** The bottom layer of the stack of cubes has just one cube at the front (so one square on the base of the plan), so we can discount answer **e**. The next row has one cube to the left, then a gap, so we can discount **a** and **d**. Looking at the back row, there is one square on the base, then a gap and another stack to the right. This must discount answer **c**, so the answer must be **b**.

9 **a** The front 'arm' of the picture has one cube sticking out beyond the rest on the left-hand side, so the plan will have one square on the base to the left, with two more squares beginning the next two rows. The back 'arm' only has two cubes so there will be two squares at the other end of the plan, on the top and middle rows. There are cubes at different levels joining these two 'arms' on the back row, but no other cubes in the second row, so the answer must be **a**.

10 **c** The bottom layer of the stack of cubes has two cubes at the front, separated by a gap (so it looks like there will be two squares separated by a space on the bottom row of the plan). The next row has cubes at various levels filling in a row of three. Looking at the top of the stack, there is a cube sticking out into the bottom row. This is on the third cube in the row, so will fit over the space between the two cubes at the front. There is a cube sticking up into the top row on the left, but no cubes behind the tallest stack, so **c** must be the correct answer.

11 **d** **rotation** – the picture is rotated 225° clockwise; **position** – the flag points towards the circle; **shading** – the shading on the flag is always black and the circle is always white.

12 **e** **rotation** – the picture is rotated 90° clockwise; **shape** – the picture forms a 'G' shape with the

tail looping around and finishing parallel to the back edge; **position** – (a) the cross in the square touches the corners, (b) the black triangle is on the first corner to the left of the open end of the shape.

13 e **rotation** – the picture is rotated 90° clockwise; **line style** – the 'corner' shape has double lines on both sides; **position** – (a) two small lines fit over the end of the arm of the corner shape to the right of the apex, (b) the sides of the small square are parallel to the arms of the corner shape; **shading** – the circle inside the small square is black and outside is white.

14 d **rotation** – the picture is rotated 270° clockwise; **position** – (a) the diamond is on the left-hand point of the outer 'U' (when it is upright) and the circle is on the right-hand point, (b) the inner 'U' is upside down inside the outer 'U', (c) the shading on the central circle is on the left when the inner 'U' is upright; **shape** – the inner 'U' loops back round on itself to join the circle in the centre.

15 b **rotation** – the picture is rotated 45° clockwise; **position** – (a) the quadrilateral is always to the right-hand side of the 'arrow' when it is upright, (b) the circle is always beneath the quadrilateral, when the 'arrow' is upright; **shape** – the arrowhead lines are approximately two thirds of the depth of the arrow; **shading** – the shading on the quadrilateral runs parallel to the base of the 'arrow'.

16 d The left-hand 'L' shape is rotated 45° forward around a vertical line, then the second 'L' sits at the back of it, fitting around the inner corner. There is then one single cube behind this second 'L', one on top at the front and one at the right-hand side.

17 c The 'L' shape rotates 90° to the left and fits over the right-hand side of the 'T' shape. The single cube then sits to the right at the front and the final cuboid rotates to stand on its end at the front as well, hiding the bar of the 'T' shape.

18 b The long cuboid drops down to form the base of the shape. The 'L' shape then hangs over the middle of this cuboid to the left. The two single cubes also sit on top of the long cuboid and the two-cube cuboid stands on its end at the back.

19 d One 'L' shape stands upside down at the front, hanging over the two-cube cuboid to the right, the other 'L' hangs over it to the left at the back. The single cube stands on top of the front 'L', and the three-cube cuboid stands on its end at the back.

20 a The 'T' shape stands on its side with the base of the 'T' pointing to the back. A three-cube cuboid sits beneath the bottom of the 'T' at the back, then followed by two-cube and three-cube cuboids on top to make a solid block. The final two-cube cuboid then stands on its end at the front.

Paper 5 (page 28)

1 a **proportion** – the largest shape is cut in half along a diagonal line, the bottom half is removed and the shape enlarges; **translation/proportion** – the two smaller shapes come together (one enlarges to fit inside the other).

2 c **rotation** – the whole picture rotates 90° clockwise; **translation** – one of the lines moves beneath the larger shape; **shading** – the direction of the shading doesn't rotate with the rest of the picture.

3 e **proportion** – the outer shape is cut in half along a horizontal line to form the smaller shapes; **number** – the number of small shapes in the first picture is the same as the number of shapes in the second picture; **shading** – following the pattern in an anticlockwise rotation in the first picture gives the order of the pattern, top to bottom, in the second picture.
Distractor: **shape** – the larger inner shape in the first picture is unimportant.

4 d **rotation** – the whole picture rotates 45° (either clockwise or anticlockwise); **position** – the layers swap position; **shading** – the shading remains in its relative position.

5 b **reflection** – the two large shapes reflect in a horizontal line; **shading** – the shading in these large shapes swaps; **position** – the small shapes shift one shape to the right, with the last shape on the right moving to the beginning.

6 c **reflection** – the complete picture is reflected in a vertical mirror line so the large triangle will be pointing to the left instead of to the right.
Distractors: **shape** – the shape doesn't change so there is always a small triangle in a larger one and a pair of parallel lines; **reflection** – some of the incorrect answers have been reflected in a horizontal mirror line.

7 e **reflection** – the complete picture is reflected in a vertical mirror line so the curved edge will be on the left-hand side instead of on the right.
Distractors: **position** – the quadrilateral doesn't change position in the box, other than with the reflection and always spans a corner of the outer shape; **reflection** – some of the incorrect answers have been reflected in a horizontal mirror line.

8 d **reflection** – the complete picture is reflected in a vertical mirror line so the top and bottom loops will curve to the right rather than to the left.
Distractors: **position** – the groups of three circles remain unaltered in the reflection; **reflection** – (a) some of the incorrect answers have been reflected in a horizontal mirror line, (b) the small triangle should reflect with the rest of the shape.

9 d **reflection** – the complete picture is reflected in a vertical mirror line so the whole picture flips to the right, and the black triangle on the left will now be on the right.
Distractors: **position** – (a) although the bottom and middle are correctly reflected in option **a**, the

top features haven't changed, (b) although the top and middle are correctly reflected in option **b**, the bottom features haven't changed; **reflection** – answer option **e** has been reflected in a horizontal mirror line.

10 **a** **reflection** – the complete picture is reflected in a vertical mirror line so the whole picture flips to the right so the arrow and all its features will now be pointing to the left.
Distractors: **shading** – the shading in the bottom three boxes is in the centre and reflected in **a**, **b** and **c**, it is not reflected in **d**, and is in the wrong box in **e**; **reflection** – (a) the black square in option **b** doesn't reflect, (b) the black arrow head in option **c** doesn't reflect; **shape** – the random reflections make the shape of the inner arrow appear to change when, in fact, it does not.

11 **d** **shape** – the letter R represents a triangle and the letter S an oval.
Distractor: shape – the black shape is unimportant.

12 **c** **line style** – the first letter represents line style: X is dashed, Y is thick and solid, Z is thin and solid; **number** – the second letter represents the number of circles: F is two, G is three.
Distractor: **number** – the number of triangles is unimportant; **shading** – the shading on these triangles is unimportant.

13 **b** **position** – the first letter represents position of the oval: L at the top of the box, M in the middle and N at the bottom; **number** – the second letter represents the total number of circles: R is three, S is four, T is two.
Distractor: **shading** – the shading on the circles is unimportant.

14 **c** **shape (1)** – the first letter represents the outer shape: R is a rectangle, S a triangle, T an oval; **shading** – the second letter represents the shading pattern: L is diagonal, M horizontal; **shape (2)** – the third letter represents the inner shape: F is a circle, G a triangle.
Distractor: **line style** – the line style of the small shape is unimportant.

15 **e** **shape** – (a) the first letter represents the second shape: S is for a square, T for a triangle, (b) the second letter represents the first shape: F is for a rectangle, G for a circle, (c) the third letter represents the third shape: X for a circle, Y for a triangle, Z for a rectangle.
Distractor: **shading** – the shading is random.

16 **d** **shape** – the shape changes on every row; **number** – an extra line is added in every column; **position** – the extra line is added following a clockwise pattern.
Distractor: **position** – the position of the lines don't change in relation to the outer box, but appear to because the shapes increase in size.

17 **e** **translation** – the shapes in the outer columns come together in the inner columns; **reflection** – the joined shape reflects in a horizontal line as it moves to the middle column; **shape** – the shape changes in each row.

18 **b** **shading** – the shading of the picture changes in each row; **rotation** – the rotation works in a pattern from top right to bottom left: the shape moves 45° clockwise moving down the grid; **shape** – the shape changes in each row (although there are no incorrect answer options involving shape).
Distractors: **shading** – the shading is the same in each row, so the position of the shading on the bottom shape will not move left to right; **rotation** – there are several incorrect answer options involving rotation.

19 **a** **number** – the number and pattern of the black bars increases working down the rows; **shape** – (a) the central shape changes between columns, (b) the small shape in the bottom right-hand corner is the same working top left to bottom right; **position** – the position of the lines is the same working top right to bottom left.
Distractor: **rotation** – none of the shapes rotates.

20 **c** **number** – the number of small lines increases moving around the outer edge of the picture in a clockwise direction; **shading/pattern** – the shading/pattern on the small shape changes between columns; **shape** – the small shapes match in diagonals from top left to bottom right; **position** – the position of the large white triangle works in a diagonal from top right to bottom left.
Distractor: **shape** – the shape of the grid makes it difficult to spot the two diagonals.

Paper 6 (page 33)

1 **e** When the net is folded to make a cube, the two squares will be next to each other. As these are folded down, the face to the right (the white circle) will join to show the cube in answer **e**.

2 **b** When the net is folded to make the cube, the top of the 'U' shape folds in to meet a black square in the chequered face. The face below the 'U' will be the single black triangle in the position shown. If the net is turned 90° anticlockwise, it is easier to see that **b** is the correct answer.

3 **b** Looking at the flat net, the side with the diagonal stripe goes to the corner between the cross and the diamond. Therefore, when the net is folded to make a cube, the top of the diagonal will be between these two faces, as in answer **b**.

4 **a** Thinking of the central faces of the net as a continuous loop, when it is folded to make the cube, the face with the black triangle will point towards the face with the horizontal stripes. The face with the white rectangle will fold down to meet the striped face on the left-hand side, so **a** must be the correct answer.

5 **b** Imagine turning the net upside down so that the bottom face is at the top, the triangle sits across the top, pointing to the right, as in option **b**. It is then easy to see that the circle will then face beneath this when the net is folded and the star will be on the right.

6 **d** Shape **d** is rotated 90° clockwise around a horizontal line.

7 **a** Shape **a** is rotated 90° clockwise around a horizontal line.

8　c　Shape **c** is rotated 90° clockwise around a horizontal line and then 90° clockwise around a vertical line.

9　e　Shape **e** is rotated 90° away from view, and then rotated 90° clockwise on its base.

10　b　Shape **b** is rotated 90° clockwise around a vertical line.

11　The pattern is revealed by overlaying the pattern of circles in the first box onto the second box, with the top-left circle being placed on the second circle from the left on the top row.

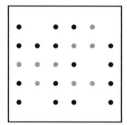

12　The pattern is revealed by overlaying the pattern of circles in the first box onto the second box, with the top-right circle being placed on the first circle in from the right on the top row.

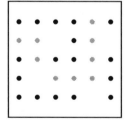

13　The pattern is revealed by overlaying the pattern of circles in the first box onto the second box. The picture rotates 90° anticlockwise with the resulting top-right circle placed on the second circle in from the right on the top row.

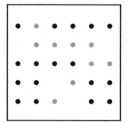

14　The pattern is revealed by overlaying the pattern of circles in the first box onto the second box. The picture rotates 90° anticlockwise and the resulting circle at the bottom right is placed on the bottom-right circle of the grid.

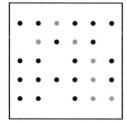

15　The pattern is revealed by overlaying the pattern of circles in the first box onto the second box. The picture rotates 90° anticlockwise with the resulting top left

circle placed on the second circle in from the left on the top row.

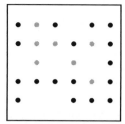

16　c　One of the three-cube cuboids lies flat with the other placed on top of it. The two single cubes then sit to the left and right.

17　a　One of the 'L'-shaped cuboids rotates 90° anticlockwise around a vertical line to face forward, the other then drops back to fit around the base. The three cubes then stack up at the front of the upright 'L'.

18　e　The 'T'-shaped cuboid rotates 90° anticlockwise around a horizontal line to stand upright. The 'L'-shaped cuboid then moves back to sit beneath the right-hand side of the 'T'. The two single cubes the sit on top of the 'T' and the two-cube cuboid moves beneath the left-hand side of the 'T'.

19　d　The 'U'-shaped cuboid stays in the position shown in **d**, and the 'L'-shaped cuboid rotates 90° clockwise around a horizontal line to sit on the back of this shape. The two single cubes then stack in front of the 'L' to complete the image.

20　b　The 'corner'-shaped cuboid stays in the position shown in **b**, and the longer 'L'-shaped cuboid rotates 90° clockwise around a horizontal line to sit on the top of this shape at the front. The smaller 'L' shape then falls on its side to the right and wraps around the bottom of the 'corner' shape (so is only visible at the front and on the right-hand side). The single cube then sits on top at the back to complete the image.

Paper 7 (page 37)

1　e　**position** – the circle is always on the shaded side of the angle, except for answer **e**.
In common/distractors: **size** – the size of the angle is random; **position** – the side of the angle the circle sits on changes randomly.

2　c　**number** – there is only one shaded circle, except for answer **c**, which has two.
In common/distractors: **shape** – the outer shapes are all different; **position** – the position of the dashed central line can be either vertical or horizontal; **shading** – the shading in the large shape is random.

3　b　**position** – the line from the centre of the long inner line always goes to a corner, except for answer **b**, where it goes to a side.
In common/distractors: **shape** – triangles vary in shape; **position** – triangles all vary in position, making it difficult to compare the size of angles.

4　e　**position** – if the square is inside the shape the line is thick and solid, if it is outside the shape it is thin and solid.

Distractors: **shape** – the larger shape/shapes are unimportant; **number** – the number of sides is unimportant; **symmetry** – there are no relationships of symmetry; **position** – the orientation of the square is unimportant.

5   **d**   **number** – each picture is made up of three lines and a square with a cross, except for answer **d**, where there are four lines.
In common/distractors: **position** – the position of the cross in the square, either touching the sides or corners, is random; **angle** – the angles made between the lines varies between the pictures.

6   **e**   **rotation** – the whole picture rotates 90° clockwise; **reflection** – the small shape reflects in a horizontal line and appears inside the large shape; **shading** – the shading moves onto the reflected shape.

7   **a**   **translation** – the lines in the left-hand picture move to make the sides of the shape in the right-hand picture; **shading** – the shading in the small shape becomes the shading inside the new shape formed, with the small shape becoming white.

8   **d**   **position** – the shapes move around using the rule: right-hand shape moves to the left, left to the middle, middle to the right.
Distractors: **shading** – the patterns and shading do not change; **shape** – the shape does not change.

9   **b**   **reflection** – the whole picture reflects in a horizontal line; **translation** – the small shape on the right moves over the cross on the left; **position** – the position of the shading reflects in the central vertical line.
Distractor: **rotation** – the cross does not rotate.

10   **c**   **rotation** – the whole picture rotates 90° clockwise.
Distractors: **shape** – the shapes do not change; **shading** – the shading rotates with the main picture.

11   **e**   The fold lines act as lines of reflection, with a separate line of reflection for each small shape. The small square folds to the right so sits at the bottom-left corner of the central square; the triangle reflects in the vertical line to its left so the right-angle sits in the bottom right-hand corner. Finally the quadrilateral at the top reflects in a horizontal line so the short edge is about one-third of the way down the picture.

12   **b**   The fold lines act as lines of reflection, with a separate line of reflection for each small shape. All the triangles reflect in the horizontal lines at the top and bottom of the square, the quadrilateral reflects in the vertical line on the left-hand side of the square. Therefore the triangles turn upside down and the quadrilateral flips left to right.

13   **d**   The fold lines act as lines of reflection, with a separate line of reflection for each small shape. The top triangle folds along the horizontal line at the top of the square, folding down like an envelope flap. The quadrilateral on the left flips to the right on the vertical line formed by the left-hand side of the square, the triangle on the right flips to the left on the vertical line formed by the right-hand side of the square.

14   **d**   The fold lines act as lines of reflection, with a separate line of reflection for each small shape. The shapes fold to meet each other so the pattern is difficult to follow but **d** is the only possible answer. The triangles at the top and bottom fold along horizontal lines so the sides toward the middle of the square form a continuous line. The triangles to the left and right fold along vertical lines with the left-hand triangle touching the central line formed by the top and bottom triangles. The triangle on the right then forms a diagonal since the top of the triangle is hidden on the top edge of the square.

15   **d**   The fold lines act as lines of reflection, with a separate line of reflection for each small shape. The triangles fold along horizontal lines: the top two fold down, so their right-angles stay on the left, but the triangles point downwards; the bottom triangle points upwards with the right-angle staying on the left. The quadrilateral on the left flips to the right on the vertical line formed by the left-hand side of the square, the square on the right flips to the left on the vertical line formed by the right-hand side of the square.

16   **e**   **reflection** – the complete picture is reflected in a vertical mirror line, so instead of the 'arrow' shape on the left of the picture pointing to the right, it will be on the right, pointing to the left.
Distractors: **position** – the two lines coming up from the base are angled to the right and parallel to each other; **number** – the picture is made up of five lines.

17   **e**   **reflection** – the complete picture is reflected in a vertical mirror line, so instead of the shaded semi-circle being on the left-hand side, it will be on the right-hand side.
Distractors: **shape** – the semi-circle doesn't change shape; **reflection** – the shapes do not reflect horizontally; **position** – the double lines do not change position.

18   **c**   **reflection** – the complete picture is reflected in a vertical mirror line, so instead of the diagonal line running from top right to bottom left, it will run from top left to bottom right.
Distractors: **shading** – the two triangles keep their original shading; **reflection** – (a) the diagonal shading reflects so runs top left to bottom right, (b) the black triangle does not reflect horizontally; **proportion** – the length of the diagonal line doesn't change.

19   **d**   **reflection** – the complete picture is reflected in a vertical mirror line, so instead of the 45° angle pointing up to the right, it will point up to the left.
Distractors: **shape** – the inner semi-circle doesn't join to the base of the outer ring; **position** – (a) the two shaded bars in the outer ring do not move up and down, (b) the two small lines don't swap, (c) the thinner short line is always on the diagonal line joined to the inner semi-circle.

20 b **reflection** – the complete picture is reflected in a vertical mirror line, so instead of the concentric circles being at the top left, they will be on the top right.
Distractors: **reflection** –there is no horizontal reflection taking place; **rotation** – the 'L' shape does not rotate so the single vertical line will be on the right; **position** – (a) the diagonal line at the bottom of the picture always goes to the white circle, (b) this line finishes at the centre of the white circle.

21 c **shape** – the letter F represents a triangle, while the letter G represents a rectangle; **shading** – R represents diagonal shading, while S represents black shading.

22 d **shape** – the first letter represents the second shape and the second letter represents the first shape: (a) the letter X represents the square, Y the triangle and Z the pentagon, (b) L represents the circle, M the hexagon and N the diamond.
Distractor: **position** – the order of the shapes makes this question more challenging.

23 e **shape** – the first letter represents the shape: R for rectangle, S for triangle, T for five-sided polygon; **number** – the second letter represents the number of circles: W for three, X for one, Y for two, Z for four.
Distractor: **shading** – the shading of the circles is random and unimportant.

24 a **shape** – the first letter represents the shape of the line at the top: F is a smooth curve, G a zig-zag, H is square-shaped; **position** – (a) the second letter represents the position of the central bar: in P the horizontal is above the small shapes, in Q it is beneath, (b) the final letter represents the number of small shapes: X is four, Y is three.
Distractor: **line style** – the line style of the central bar shape is random.

25 b **shape** – the first letter represents two outer shapes: F is a smooth curve, G has straight lines formed in a point, H is square-shaped; **line style** – the second letter represents the line style of these shapes: in L it is thick and solid, in M fine and solid, and in N dashed; **shading** – the final letter represents how the small circles inside these shapes are shaded: R is both black, S is one black and one white, T is both white.
Distractor: **position** – the circles placed around the edges and outside the two shapes are unimportant.

26 b **number** – an extra circle is added in each box, working left to right across the sequence; **rotation** – (a) the triangles rotate anticlockwise around the box, (b) the quarter circles rotate clockwise by one quarter of a circle.

27 b **position** – the square moves upwards in a diagonal from bottom left to top right, working left to right across the sequence; **rotation** – the 'flag' shape rotates 45° clockwise; **number** – the number of concentric circles works in a symmetrical pattern from the centre: one, then two, then three.

28 e **rotation** – the solid line rotates 45° anticlockwise around the box, working left to right across the sequence; **reflection** – (a) the dashed line alternates between bottom left and top right, reflecting in a diagonal line, (b) the triangle works in repeating sequence in alternate boxes: bottom left to top right, reflecting in the central diagonal each time, (c) the curved shape also moves in a repeating sequence in alternate boxes, reflecting in the central diagonal; **shading** – the shading on the curve alternates between white and black.

29 d **position** – (a) the circle moves one side anticlockwise around the box in the top row of the sequence, (b) the arrow moves one corner anticlockwise around the box in the bottom row of the sequence, mirroring the direction of the circle, (c) the black circle moves from one side of the triangle to the other as the shapes move around the box; **rotation** – the triangle feature rotates 90° anticlockwise as it moves around the box; **number** – one extra triangle is added in each box.

30 a **position** – the small square moves one corner anticlockwise, moving up and down across the sequence; **rotation** – (a) the black triangle in the small square rotates 90° anticlockwise inside the square, moving up and down across the sequence, (b) the corner shape with the arrow heads rotates 90° clockwise within the box, moving left to right and up and down across the sequence.
Distractors: the position of the missing box makes it difficult to follow the direction of the sequence; **position** – the position of the shaded arrow heads doesn't change, but the split sequence makes this difficult to follow.

Paper 8 (page 44)

1 d When the net is folded to make the cube, the face with the star will wrap around to be next to the four circles on the face at the opposite end of the net. The face with the two concentric squares will then fold over to join them, in the position shown in answer option **d**.

2 c When the net is folded to make the cube, the chequered face and the face with the diamond will remain next to each other. The face with the triangle will then fold down into the position shown in answer option **c**.

3 c When the net is folded to make the cube, the face with the horizontal stripe will be next to the face with the diagonal stripe. When the face with the striped triangle is folded down to meet the diagonal, the corner of the triangle will be next to one end of the stripe, as in option **c**.

4 e When the net is folded to make the cube, the blank face will wrap around to join the face with the white circle. The black semi-circle will then fold down to meet the blank face, as in option **e**.

5 c When the net is folded to make the 3D shape in the position shown, the hexagon will be on the face to the left of the blank face. The triangle on the face above will be pointing downwards, as in option **c**.

6   **a, b, d**   **a** sits at the top of the hexagon, next to **d**, then **b** fills in the remaining left-hand side of the shape.

7   **b, d, e**   **e** rotates 90° anticlockwise and sits on top of **b** to make a rectangle, then **d** is placed beneath **a** to complete the shape.

8   **a, c, e**   **e** rotates 90° clockwise to join with **a**, then **c** rotates 45° anticlockwise, fitting in the space at the top of **a** to complete the shape.

9   **a, b, c**   **c** rotates 45° clockwise, **b** rotates 45° anticlockwise and sits beneath **c** to form the bottom portion of the shape, **a** then rotates 135° anticlockwise, fitting in the space at the top left to complete the shape.

10   **b, c, e**   **e** rotates 180° to sit on top of **b**, **c** then rotates 90° clockwise to sit on top of **e** and complete the shape.